Solar Power

How to Use a Solar Power
System and Other Renewable
Sources to Cut Your Electric
Bill to Zero

by Joseph Halley

Table of Contents

Disclaimer

While all attempts have been made to verify the information provided in this book, the author does not assume any responsibility for errors, omissions, or contrary interpretations of the subject matter contained within. The information provided in this book is for educational and entertainment purposes only. The reader is responsible for his or her own actions and the author does not accept any responsibilities for any liabilities or damages, real or perceived, resulting from the use of this information.

Introduction

Electricity is absolutely essential to modern civilization and is growing in importance in many areas of the world as many countries in South America, Africa and Southeast Asia continue to develop and become more industrialized. Electricity is the source of our light, our entertainment, our heat and usually our escape from the heat. Most of us cannot imagine our lives without it.

Yet, it comes with a cost. First, there is the fact that it most certainly is not free. Second, how much it costs often depends on the actions of countries far removed from our own as the source of the electricity we use to charge our smartphones is very often oil or some other fossil fuel traded on the global market. The fact is, OPEC affects how big your electric bill is.

Third, there is the environmental cost as fossil fuels and the processes to remove them from the earth release pollutants into the air that have been linked to various cancers and respiratory issues.

There are certain cities in the world (Hong Kong, Los Angeles) where the air is so bad that the citizens often are told to stay indoors because the air is so bad it can be difficult to breathe.

In response to these financial and environmental pressures, a movement has arisen that aims to transition the world's energy economy away from fossil fuels to renewables like solar and wind power.

Most of the headlines focus on massive government sponsored projects like wind and solar farms that take up hundreds of acres of land and hundreds of millions of dollars. What can the average citizen do to contribute to the transition towards renewable resources compared with projects of such scale?

Fortunately, quite a bit. There are a number of different systems designed for residential use that will allow nearly anyone to reduce their electric bill and carbon footprint.

In fact, some have even had so much success that instead of getting a bill from the local power company, they now get a check.

This guide will walk you through some of the various systems – solar, wind, hydro, and more – that will help you decrease your reliance on the traditional power grid as well as make your home more self-sufficient and environmentally friendly.

We will take a look at the different available systems and give you an idea of how much you can expect to spend and how long it should take to get a return on your investment.

Chapter 1 – Getting Started

First Things First – Checking the Rules

No matter where you live, there are rules governing what you can and cannot do with your home. These rules will vary depending on where you live and tend to be more restrictive the less rural your neighborhood.

There are good reasons for this in many instances and some of them are fairly obvious. For example, it should be pretty clear that you are not going to be able to set up a 100 ft. tower with a wind turbine on top in the middle of a dense residential area.

You can start by checking your local township or city's website to see if they have information on what codes and other regulations apply to your area. The amount of information on such sites varies considerably of course and you may find yourself stuck with either information so general it is of little help or stuck in a swamp of minutiae that would cross a lawyer's eyes.

Should your efforts on the web prove unfruitful, try to meet with the local inspector who can walk you through the relevant codes and how they apply to you. He can also recommend good contractors who have experience installing solar power and other systems.

Don't forget to check your homeowner's association rules as well, if applicable. Many of these kinds of communities have rather bizarre and arcane rules that you might run afoul of if you start putting solar ovens in your backyard. If there are no rules that seem to apply to your solar farm or plans for a living roof, it may be a good idea to bring that up at the next association meeting.

On the positive side, it is also worth looking into any subsidies and tax credits. The government has taken a very active interest in promoting renewable energy sources and as such has offered numerous incentives to encourage people to adopt energy efficient and renewable technologies.

Before deciding on what sort of system would suit your needs, it is worth investigating the various financial rewards made available by federal, state, and local governments.

Next Step – Reduce

One more thing that you should do before spending any money on a renewable energy system is reduce how much energy your household consumes. Naturally, this is easier to do if you are building a new home.

Plans can be made from the beginning for extra, energy efficient windows and appliances, high quality shingles, extra insulation and the like.

You can even plan from the beginning for geothermal heating (more on that later) and solar panel hook ups, minimizing conversion and retrofitting costs.

However, most of us are in comfortable homes and have neither the funds or desire to move. Even in these situations appliances, roofs, and lightbulbs need to be replaced.

The lowest hanging fruit here is your lights. If you are still using incandescent lightbulbs, a significant way you can reduce your energy consumption is to switch to LED lighting.

The energy consumed is up to ten times less than incandescent and roughly half of what is consumed by CFL bulbs. That also means that your operating cost is roughly ten times less.

It is true that LEDs cost considerably more but given the energy savings and the fact that they have a considerably longer operating life than your average incandescent bulb, they will pay for themselves fairly quickly.

Beyond this simple change, you can replace your appliances with more energy efficient models, further reducing your energy consumption. Considering that appliances in general are not cheap and the most energy efficient models are more expensive still, I would recommend waiting until your current units need to be replaced.

However, whether you choose to wait or you have the extra money to move ahead immediately, make sure that you research thoroughly to find the best models before buying. Also take into account if it is possible to find someone locally to service the unit.

If you want your energy savings to be cost effective, it will not pay if you have to get parts or service from hundreds of miles away to keep a refrigerator up and running.

The same situation applies to your windows. They are absolutely not cheap, yet they break or their seals degrade and eventually need to be replaced. When you do need to replace them, go to www.energystar.gov where you can find a list of the most energy efficient windows and appliances for the current year.

Doing this, you can minimize the heat loss from your home during the winter and that same improved insulation will also help keep your home cooler in the summer.

One other big ticket item that no one wants to replace but is necessary at some point is the roof. Should that time come, there are new technologies that will again reduce your heating and cooling needs.

There are now shingles that are designed to reflect light and therefore heat from your home. Not long ago, such shingles were necessarily white as that color reflects more light than any other.

However, recent advancements have allowed companies to create highly reflective, energy efficient shingles in many other colors, making them more appealing to those who want to be more energy efficient but don't want to stick out like a sore thumb.

And if you are thinking about how that is likely to hurt you in the winter, the snow will cover those shingles and provide plenty of insulation.

Finally, if you have a significant sum of money lying around and have been thinking about a remodel, there are a couple of options to keep in mind. One is that instead of just replacing your windows, you could put in extra windows, specifically windows on the south side of your home.

This will allow sunlight and heat directly into your home during the morning. When the sun starts to go down, the furniture and carpet that are in the room will begin releasing the heat that they have absorbed through the day, helping keeping your home warn throughout the evening and into the night.

A related option is to add a sun room. Again, if possible, place the room on the south side of the home to make sure that you are getting lots of direct sunlight which of course is free heat.

Naturally, this will not always be possible. Wherever you choose to build it though, make sure that there are lots of windows to allow the light and heat to come streaming in. Air flow is important here as you want that excess heat distributed to other areas of your home and not just trapped in one area.

Chapter 2 – To Grid or Not to Grid

One large motivator for transitioning your home to a solar power or other renewable system is to be able to live off the grid, independent of the larger energy infrastructure. While this is certainly doable in many areas, it is not always feasible and will usually cost more.

The reason an off-grid system costs more is that you will need to hook up a battery back up to store energy generated when the sun is shining and the wind is blowing to make up for those times when it is not.

You'll need the batteries, an inverter to convert it the electricity stored in them to standard household electricity and likely a backup, gasoline powered generator for emergencies. And of course all the space to store that equipment.

There are also many climates and living situations where you may not be able to generate enough electricity with renewable means. If you live in an area with extensive cloud cover and unreliable winds, chances are, renewables will only be a supplement for you and an off-grid system will be impossible.

Urban settings also negate the possibility of wind power altogether and depending on regulations could possibly your solar power system as well, to say nothing of a lack space for all the equipment discussed above.

Rural areas will have an easier time, both due to more available space and generally lighter regulations governing property use. Depending on how rural you are, for example if you are building far away from any pre-existing power lines, it may actually be cheaper to build an off-grid system that it is pay the power company to run a line out to your property.

It will also likely be more reliable in a storm as well since rural areas are usually at the bottom of the list when it comes to repairing damaged power lines.

A system connected to the grid has the advantage of not needing all of the batteries and other equipment to keeping things going during the rainy season. It also has the distinct advantage of being a possible way to help recoup your costs.

If you live in a state that has net-metering laws in place (which is most of them) your meter can actually run backwards if you are generating more electricity than you are using. Many people actually get a check from the power company at the end of the year.

It is not much generally, on the order of a few hundred dollars but when combined with the fact that you also are not paying for any electricity, the savings quickly add up and your system can begin paying for itself sooner rather than later.

The downside to this situation is that in the event of a blackout due to whether, energy shortages or a downed power line that causes a chain reaction like what happened in the eastern part of the United States back in 2003, you will only have the power that can be generated by your system when conditions are right.

You will still be in better shape than your neighbors in such a situation but if your goal is 100% reliability, then you will still need at least a generator backup for the essentials like the refrigerator and a few houselights.

My personal recommendation would be to start out with a system that is connected to the grid and then gradually add the necessary back up equipment that will allow you to go off-grid if need be. By going this route, you can start off saving money on your electric bill with a relatively small investment and gradually build up your system to where you are making money.

At that point, you start pocketing that extra cash and use it to help fund your batteries and generators until you have a system that will stand up to anything short of a nuclear blast or zombie apocalypse.

Now that we have taken a brief look at the biggest ways you can reduce your energy consumption and started thinking about whether or not it is a good idea to be connected to the power grid, it is time to get into the interesting parts, namely actually talking about solar power and other renewable systems that can be installed in your own home, allowing you to generate you own electricity and potentially moving off the grid if that is your goal.

Chapter 3 – Solar Power

Energy from the sun is literally all around us. It strikes our planet constantly, an energy source that will continue to be available to everyone on the planet for roughly the next 5 billion years when our star will become a red giant and wind up destroying the earth.

Until then though, the giant fusion reactor that is the sun is producing vast amounts of energy that so far in our history has largely gone untapped. Only in the last handful of decades have we begun to make use of this gigantic, free and very clean source of energy.

How do Solar Panels Work?

Solar cells are made up of a number of tiny cells, called photovoltaic cells. These cells are largely made up of two layers of silicon, one with an excess of electrons, one with a shortage.

When the photons from sunlight hit the cell, electrons are knocked loose and flow through your system to the other side.

That electron flow is literally the flow of electricity and your loads (batteries, light bulbs, a 55 inch LED TV) are from an electrical perspective "between" the two layers.

This process in not yet terribly efficient but it is getting better. Once, the most efficient systems could only convert 10% of the sunlight received into electricity. Now, commercial systems can do twice that and experimental systems are reaching rates as high as 46%.

Just how much energy is there in sunlight? There are conflicting claims on this one but the most recent data indicates that one square meter of earth receives an amount of sunlight equivalent to one kilowatt hour of electricity.

By itself, this is not much power, but multiply it by ten and it becomes enough to power a small home. An area the size of a swimming pool with give you enough power in the course of the day to light up a number of homes at once.

Advantages and Disadvantages

There are of course, certain obvious objections to the effectiveness of solar panels. The most obvious is that any time it snows, your panels will be blocked and generate no electricity. While the snow is certainly a difficulty, it is not insurmountable.

One can simply push the snow off with a simple snow rake for example, which should probably be done any way to prevent the buildup of damaging ice dams on the edge of your roof. While this can certainly be cumbersome, when you realize just how much money you are saving per week you will likely find that it is worth the upkeep.

Also we would be remiss if we forgot to mention the single most pervasive case against solar energy. Namely, that it only generates electricity when the sun is up and not hiding behind a cloud.

Obviously, there is quite a bit of truth here but the efficiency of solar panels have increased so much in the last serval years it is highly likely that you will generate not only enough electricity for your daytime needs, but for the night as well.

You will of course need to acquire a battery system to store the excess electricity generated by the panel to make sure that the fridge stays cold throughout the night.

And with all the advancements that have been made in the field of solar energy, some kinds of photovoltaic cells are efficient enough that they will even work during cloud cover, meaning that unless you live in the arctic circle where it can be completely dark for many days or even months at a time depending on how close you are to the pole, making enough electricity to support, or at least supplement your home should not be a problem.

A related concern is the buildup of dirt, leaves and other debris on top of most buildings' roofs. Over time, this could be a concern, but normally there is more than enough wind and rain to keep your panels clear enough to get plenty of sun.

Another clear advantage of a solar power system is the fact that it is very scalable. It is a relatively simple matter to begin experimenting with just a few panels, perhaps enough to charge your batter-operated lawn equipment or you might place a few on the shed in your backyard to power your workshop and avoid running a separate power line out to it.

Doing this will let you test out just how many panels you might actually need for the whole home, helping you save money in the long run.

Also, if you do not have a large amount of cash to buy a whole system outright, it is possible to add a few panels at a time, eventually building to the point where you have enough power for the whole home. And the money you save with each upgrade can fund the next purchase.

By contrast, the other power generation systems discussed below (wind and micro-hydroelectric) are one-time, high dollar purchases that are difficult or impossible to scale.

Solar has an additional advantage in that, due to its relatively low profile and ability to be placed almost anywhere, it is the least likely to be significantly limited by regulations and building codes.

Once, solar panels were these huge, highly visible things that sat on top of your roof. Even for the most environmentally conscious, with a few tens of thousands of dollars to spare, they were a tough sell. Over the last decade though, they have gotten more efficient, cheaper, easier to install, and perhaps more importantly, smaller.

Some companies have been producing panels that are all black, helping them to blend in better with the black shingles that are normally on homes.

New technologies have improved on this as well, particularly the advent of thin-film solar cells which can now achieve rates of efficiency (energy received vs. electricity generated) upward of 19%, an almost hundred percent improvement over that standard efficiency just a few years ago.

Thin film solar cells have allowed for the creation of solar shingles. These are panels that can fit right over your existing shingles are even direct to the tar paper in place of shingles, with the only side effect being a slightly shiny, purple tint to the roof.

They can also generate roughly 12 watts of power per square foot, meaning that even a small, single story ranch home would be able to generate a fair amount of electricity if its roof were made up of these shingles.

And these are far from the only new, smaller, less obvious kinds of solar panels on the market these days. Different companies are making solar panels that bend, fold and even peel and stick, allowing you to harness the sun's natural energy virtually anywhere on your property using almost any surface.

A single panel for any of these can generate a hundred watts of electricity, and will generally only set you back $200 or so, allowing you to get started building your system for a relatively small amount of money.

When it comes to figuring out the total cost of your system, it can get a bit tricky. Once you figure out what your energy needs are, you can roughly figure that your system will cost $1-2 a watt just for the panels.

That means that the average home using 10,000 watts of electricity will cost from $10,000-20,000.

Considering that unless you are setting up a completely off-grid system and are very much into do-it-yourself projects you will also need to pay for professional installation, the cost can as much as double (or more; get multiple quotes whenever possible).

This is where you need to remember those government incentives for moving to a renewable system.

For example, an inn in New Jersey recently installed a full solar panel system at the show stopping cost of $75,000. First, the average homeowner will likely not need as substantial a system as a business.

Second, that business was able to receive a federal tax credit that was nearly equal to a third of the installation cost. Add in other state incentives, and the final cost was much less than $75,000. They were also able to sell excess electricity and so were able to start recouping the costs sooner rather than later.

At the end of the day though, costs are continually dropping. Back in 2008, a system could cost as much as $7 per watt to have installed. Since then that cost has dropped by as much as half and considering the increased pace of advancement that cost will continue to drop and may soon reach levels that are reasonable for most middle-class families.

Do not forget that the lower the cost gets, the shorter the payoff period is, dropping from 20 years in 2008 to something closer to 10-15 years or less now.

Another, little talked about use of solar power is in heating. Panels can be installed that instead of being filled with photovoltaics are filled with pipes.

Water is pumped up through those pipes where the sun heats it before it goes back down into the home in a radiant heat system, not unlike the boiler-based radiant heat systems still present in many older homes.

In fact, since it is essentially the same system with a different source of energy, this becomes a very simple way to integrate renewable green energy into even an old home. As you might imagine, this also is not cheap, but as it can reduce heating costs considerably, it too will pay for itself in time.

Chapter 4 – Wind and Everything Else

Wind

Wind power has been grabbing headlines and skylines in many places around the world lately. Any drive through the American Plains States will reveal vast vistas filled with massive, commercial grade windmills meant to take the place of fossil fuels as a primary source of power.

However, while these giant windfarms have grown in notoriety, residential wind power remains a very small industry.

There are a few important reasons for this. One is that wind turbines necessarily have far more moving parts and thus maintenance than any solar power system that you are likely to have.

This is one of the things is actually holding it back from being the commercial force many expect it to be as the maintenance is expensive and is one of the things that drives the necessity of government subsidies to make it competitive.

We also have already mentioned that for wind power to be effective, the turbine needs be on top of a hundred foot tower in order to see a minimum average 9mph wind speed, a tower that will not be allowed in nearly any residential neighborhood.

Add in the large start-up cost, the fact that the wind is not always blowing and if there are a lot of trees or buildings nearby, they will slow down wind speeds even at a hundred feet, wind may not seem all that attractive.

All of those things considered, wind can still have a role to play in a rural renewable energy system. Placing a single turbine on your property can by itself generate a large amount of electricity, up to 10kw, which is enough to power most small homes.

Combine this with the fact that the wind blows at night and when the clouds come to block the sun and you have a reliable and plentiful source of electricity to back up the primary solar power system.

And again, if you are hooked up to the grid, you can pocket even more money at the end of the year. It also isn't necessary to jump to the 10kw system either. If you would rather plan on using wind solely as a supplement, you can actually buy a smaller and thus much cheaper model.

So how much does a winds system cost? Again, this depends greatly the size of the system you want to install. In 2013, a 1kw fan could cost as much as $9000 without the batteries while a full 10kw system with everything needed could run as high as $39,000. But again, that was before tax credits which could reduce the final cost to less than half that.

A web search today shows that you can get the turbine itself for as low as $1 a watt but it is important to keep in mind that does not include installation costs or the tower for that matter, which will quickly boost the cost to something that will make you thankful for those credits and incentives.

Geothermal Heating and Cooling

One of the more interesting and popular, but least talked about renewable energy sources is geothermal. Residential geothermal does not generate electricity but it is excellent as a supplemental source of heating and cooling.

The way it works is really very simple. A PVC pipe is sent well below the surface to where the ground generally remains a steady 50F regardless of what is going on with weather above ground.

In the winter this water is heated to roughly 50F and then is brought up to the heating system where it effectively jumpstarts it, meaning that the gas or electric heating system doesn't have to work nearly as hard to keep the home at a comfortable temperature during the cold months.

In warmer weather, the water running down those pipes is cooled from whatever the temperature is on the surface to approximately 50F and then brought back to the surface where it cools the home and may eliminate the need for an air conditioning system altogether.

This system is expensive to install and it will rarely replace conventional heating. However, if you combine it with a solar or wind supplied electric heating system, it may be possible to ditch the gas or oil furnace all together.

For this system, the payoff period is as positively affected by tax credits as other renewable sources. In 2009, Popular Mechanics ran a story on geothermal heating.

During that article, Popular Mechanics showed that based on a $30,000 system and an $11,000 tax credit, the payoff could happen as soon as ten years.

This period of course will vary by considerable margins depending on the price of natural gas, propane, fuel oil, or whatever your particular source of heat is in your home.

Micro-hydroelectric

Another, less well known system for renewable energy is micro-hydroelectric power. Working on the same principle as the large dams that provide a substantial amount of power to much of the nation, companies have begun deploying smaller systems that can be installed in streams and generate enough power by themselves to keep a small home going.

The chief advantage of this system is that unless it has gotten cold enough for long enough, the water is always flowing, day or night, rain or shine, making it perhaps the most reliable renewable source of electricity. The down sides are both geographic and regulatory.

Geographic in the sense that if you are not fortunate enough to have a stream running through your property, you will not be able to take advantage of this technology.

And regulations governing water use may prevent you installing such a system even if it is technically possible.

A final downside is that a micro-hydro system can be extremely expensive as compared with others, up to six times more, though it can last a very long time and still pay for itself in the long run. As with the other renewable systems here discussed, it is likely that as this technology advances, it will come down in cost.

Conclusion

As you can see, there are any number of ways to transition your home heating and electrical needs from the fossil fuel dependent power grid to a more renewable and ultimately cheaper independent power system.

While solar is in many ways the simplest and most versatile, there are plenty of other options to reduce both your electricity use and your dependency on heating from other sources such as natural gas and propane.

Given the varying demands of different climates and variations in weather patterns from one year to the next, exploring two or more of the systems discussed in this guide will help you decide on the system that is the most cost effective and reliable for your particular situation.

Exploring other options not discussed in this guide such as solar ovens and heat pumps can further help you reduce your dependency on the larger power grid, helping to move yourself, and bit by bit, your country to energy independence and a cleaner environment.

And while it is true that many of the technologies talked about in this guide are beyond the reach of many people, due to living in an urban setting or simply because the upfront costs are too high, using some of the tips at the beginning of this guide like switching to LED lighting and gradually upgrading to more energy efficient appliances will allow anyone to join in the renewable revolution and cut those energy bills.

Whatever system you decide to implement and whether you decide to start small with a few light bulbs or jump right in with a full set of solar shingles and a wind turbine.

References

Information on energy standards and what products to the best job of meeting and exceeding them.

www.energystar.gov

Guide to the advantages and disadvantages of different kinds of lighting.

www.designrecycleinc.com/led%20comp%20chart.html

Tons of information on solar power developments.

www.solarpowerauthority.com

Tons of information on how to save energy and generate some of your own.

www.energy.gov/energysaver/energy-saver

www.ingramcontent.com/pod-product-compliance
Lightning Source LLC
Chambersburg PA
CBHW071833200526
45169CB00018B/1422